CHAPTER 3

ALTERNATING CURRENT GENERATORS

LEARNING OBJECTIVES

Upon completion of this chapter, you will be able to:

1. Describe the principle of magnetic induction as it applies to ac generators.

2. Describe the differences between the two basic types of ac generators.

3. List the advantages and disadvantages of the two types of ac generators.

4. Describe exciter generators within alternators; discuss construction and purpose.

5. Compare the types of rotors used in ac generators, and the applications of each type to different prime movers.

6. Explain the factors that determine the maximum power output of an ac generator, and the effect of these factors in rating generators.

7. Explain the operation of multiphase ac generators and compare with single-phase.

8. Describe the relationships between the individual output and resultant vectorial sum voltages in multiphase generators.

9. Explain, using diagrams, the different methods of connecting three-phase alternators and transformers.

10. List the factors that determine the frequency and voltage of the alternator output.

11. Explain the terms voltage control and voltage regulation in ac generators, and list the factors that affect each quantity.

12. Describe the purpose and procedure of parallel generator operation.

INTRODUCTION

Most of the electrical power used aboard Navy ships and aircraft as well as in civilian applications is ac. As a result, the ac generator is the most important means of producing electrical power. Ac generators, generally called alternators, vary greatly in size depending upon the load to which they supply power. For example, the alternators in use at hydroelectric plants, such as Hoover Dam, are tremendous in size, generating thousands of kilowatts at very high voltage levels. Another example is the alternator in a typical automobile, which is very small by comparison. It weighs only a few pounds and produces between 100 and 200 watts of power, usually at a potential of 12 volts.

Many of the terms and principles covered in this chapter will be familiar to you. They are the same as those covered in the chapter on dc generators. You are encouraged to refer back, as needed, and to refer to any other source that will help you master the subject of this chapter. No one source meets the complete needs of everyone.

BASIC AC GENERATORS

Regardless of size, all electrical generators, whether dc or ac, depend upon the principle of magnetic induction. An emf is induced in a coil as a result of (1) a coil cutting through a magnetic field, or (2) a magnetic field cutting through a coil. As long as there is relative motion between a conductor and a magnetic field, a voltage will be induced in the conductor. That part of a generator that produces the magnetic field is called the field. That part in which the voltage is induced is called the armature. For relative motion to take place between the conductor and the magnetic field, all generators must have two mechanical parts — a rotor and a stator. The ROTor is the part that ROTates; the STATor is the part that remains STATionary. In a dc generator, the armature is always the rotor. In alternators, the armature may be either the rotor or stator.

Q1. Magnetic induction occurs when there is relative motion between what two elements?

ROTATING-ARMATURE ALTERNATORS

The rotating-armature alternator is similar in construction to the dc generator in that the armature rotates in a stationary magnetic field as shown in figure 3-1, view A. In the dc generator, the emf generated in the armature windings is converted from ac to dc by means of the commutator. In the alternator, the generated ac is brought to the load unchanged by means of slip rings. The rotating armature is found only in alternators of low power rating and generally is not used to supply electric power in large quantities.

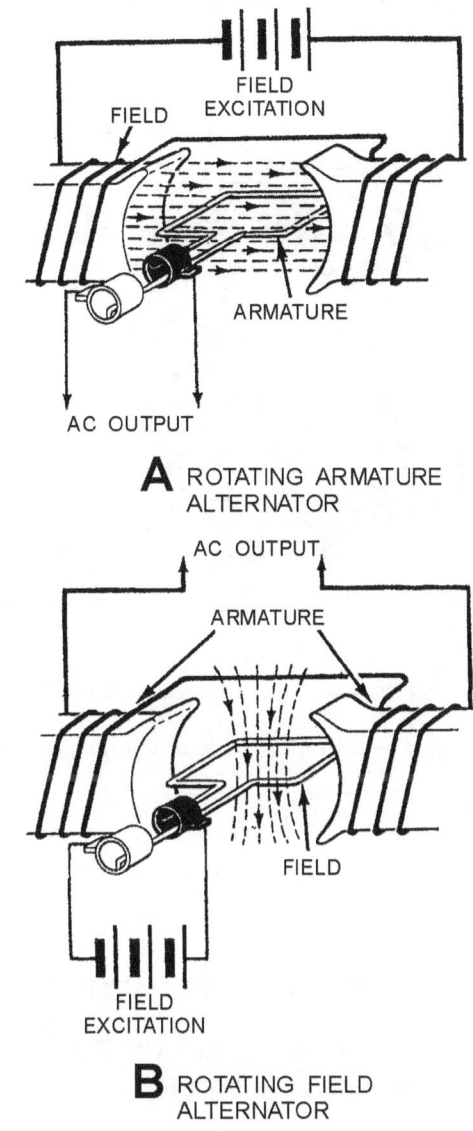

FIELD EXCITATION

FIELD

ARMATURE

AC OUTPUT

A ROTATING ARMATURE ALTERNATOR

AC OUTPUT

ARMATURE

FIELD

FIELD EXCITATION

B ROTATING FIELD ALTERNATOR

Figure 3-1.—Types of ac generators.

ROTATING-FIELD ALTERNATORS

The rotating-field alternator has a stationary armature winding and a rotating-field winding as shown in figure 3-1, view B The advantage of having a stationary armature winding is that the generated voltage can be connected directly to the load.

A rotating armature requires slip rings and brushes to conduct the current from the armature to the load. The armature, brushes, and slip rings are difficult to insulate, and arc-overs and short circuits can result at high voltages. For this reason, high-voltage alternators are usually of the rotating-field type. Since the voltage applied to the rotating field is low voltage dc, the problem of high voltage arc-over at the slip rings does not exist.

The stationary armature, or stator, of this type of alternator holds the windings that are cut by the rotating magnetic field. The voltage generated in the armature as a result of this cutting action is the ac power that will be applied to the load.

The stators of all rotating-field alternators are about the same. The stator consists of a laminated iron core with the armature windings embedded in this core as shown in figure 3-2. The core is secured to the stator frame.

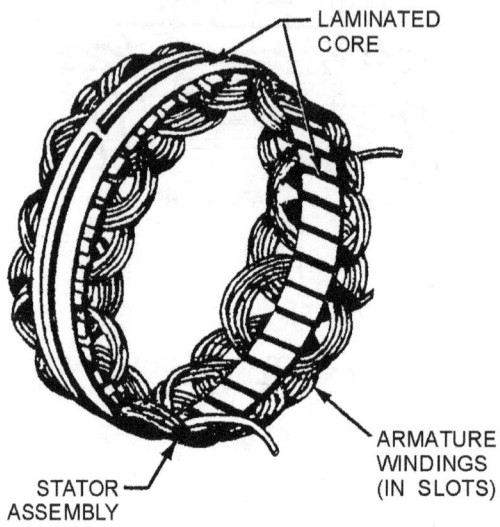

Figure 3-2.—Stationary armature windings.

Q2. What is the part of an alternator in which the output voltage is generated?

Q3. What are the two basic types of alternators?

Q4. What is the main advantage of the rotating field alternator?

PRACTICAL ALTERNATORS

The alternators described so far in this chapter are ELEMENTARY in nature; they are seldom used except as examples to aid in understanding practical alternators.

The remainder of this chapter will relate the principles of the elementary alternator to the alternators actually in use in the civilian community, as well as aboard Navy ships and aircraft. The following paragraphs in this chapter will introduce such concepts as prime movers, field excitation, armature characteristics and limitations, single-phase and polyphase alternators, controls, regulation, and parallel operation.

FUNCTIONS OF ALTERNATOR COMPONENTS

A typical rotating-field ac generator consists of an alternator and a smaller dc generator built into a single unit. The output of the alternator section supplies alternating voltage to the load. The only purpose for the dc exciter generator is to supply the direct current required to maintain the alternator field. This dc generator is referred to as the exciter. A typical alternator is shown in figure 3-3, view A; figure 3-3, view B, is a simplified schematic of the generator.

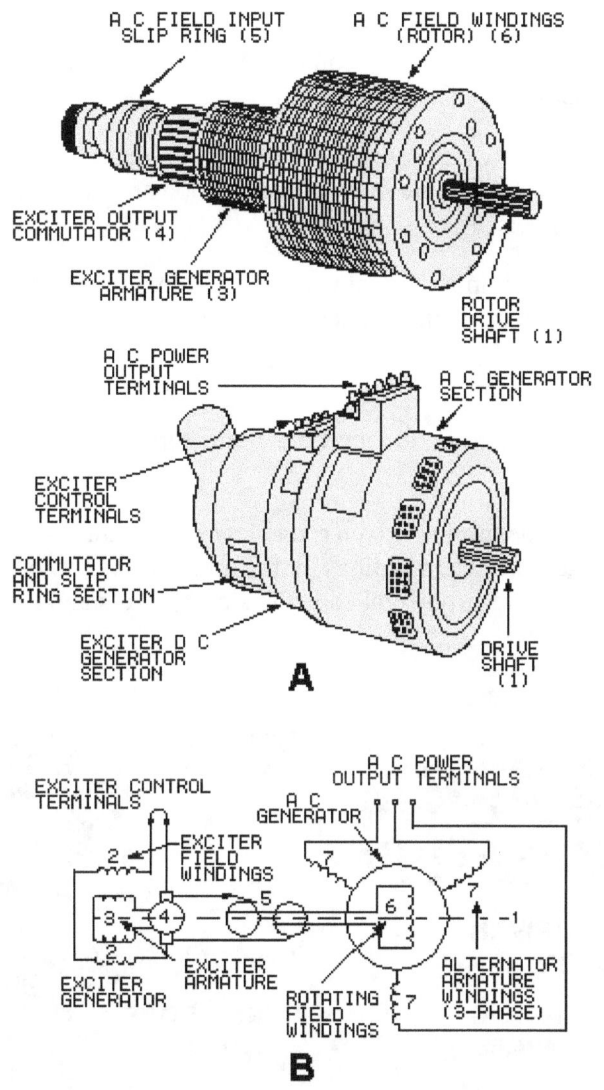

Figure 3-3.—Ac generator pictorial and schematic drawings.

The exciter is a dc, shunt-wound, self-excited generator. The exciter shunt field (2) creates an area of intense magnetic flux between its poles. When the exciter armature (3) is rotated in the exciter-field flux, voltage is induced in the exciter armature windings. The output from the exciter commutator (4) is connected through brushes and slip rings (5) to the alternator field. Since this is direct current already converted by the exciter commutator, the current always flows in one direction through the alternator field (6). Thus, a fixed-polarity magnetic field is maintained at all times in the alternator field windings. When the alternator field is rotated, its magnetic flux is passed through and across the alternator armature windings (7).

The armature is wound for a three-phase output, which will be covered later in this chapter. Remember, a voltage is induced in a conductor if it is stationary and a magnetic field is passed across the conductor, the same as if the field is stationary and the conductor is moved. The alternating voltage in the ac generator armature windings is connected through fixed terminals to the ac load.

Q5. Most large alternators have a small dc generator built into them. What is its purpose?

PRIME MOVERS

All generators, large and small, ac and dc, require a source of mechanical power to turn their rotors. This source of mechanical energy is called a prime mover.

Prime movers are divided into two classes for generators-high-speed and low-speed. Steam and gas turbines are high-speed prime movers, while internal-combustion engines, water, and electric motors are considered low-speed prime movers.

The type of prime mover plays an important part in the design of alternators since the speed at which the rotor is turned determines certain characteristics of alternator construction and operation.

ALTERNATOR ROTORS

There are two types of rotors used in rotating-field alternators. They are called the turbine-driven and salient-pole rotors.

As you may have guessed, the turbine-driven rotor shown in figure 3-4, view A, is used when the prime mover is a high-speed turbine. The windings in the turbine-driven rotor are arranged to form two or four distinct poles. The windings are firmly embedded in slots to withstand the tremendous centrifugal forces encountered at high speeds.

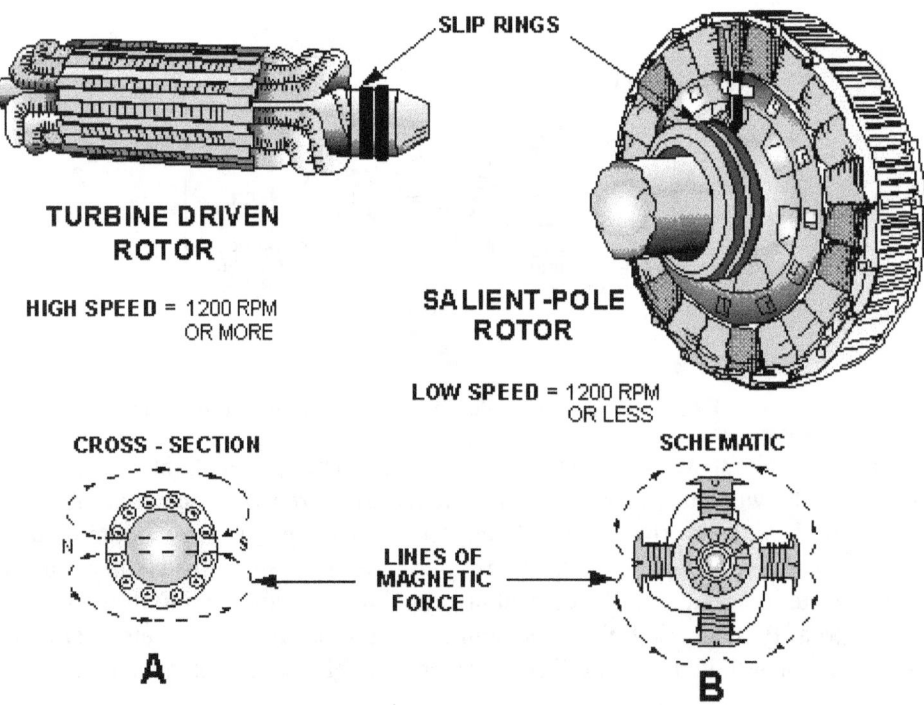

Figure 3-4.—Types of rotors used in alternators.

The salient-pole rotor shown in figure 3-4, view B, is used in low-speed alternators. The salient-pole rotor often consists of several separately wound pole pieces, bolted to the frame of the rotor.

If you could compare the physical size of the two types of rotors with the same electrical characteristics, you would see that the salient-pole rotor would have a greater diameter. At the same number of revolutions per minute, it has a greater centrifugal force than does the turbine-driven rotor. To reduce this force to a safe level so that the windings will not be thrown out of the machine, the salient pole is used only in low-speed designs.

ALTERNATOR CHARACTERISTICS AND LIMITATIONS

Alternators are rated according to the voltage they are designed to produce and the maximum current they are capable of providing. The maximum current that can be supplied by an alternator depends upon the maximum heating loss that can be sustained in the armature. This heating loss (which is an I^2R power loss) acts to heat the conductors, and if excessive, destroys the insulation. Thus, alternators are rated in terms of this current and in terms of the voltage output — the alternator rating in small units is in volt-amperes; in large units it is kilovolt-amperes.

When an alternator leaves the factory, it is already destined to do a very specific job. The speed at which it is designed to rotate, the voltage it will produce, the current limits, and other operating characteristics are built in. This information is usually stamped on a nameplate on the case so that the user will know the limitations.

Q6. How are alternators usually rated?

Q7. What type of prime mover requires a specially designed high-speed alternator?

Q8. Salient-pole rotors may be used in alternators driven by what types of prime movers?

SINGLE-PHASE ALTERNATORS

A generator that produces a single, continuously alternating voltage is known as a SINGLE-PHASE alternator. All of the alternators that have been discussed so far fit this definition. The stator (armature) windings are connected in series. The individual voltages, therefore, add to produce a single-phase ac voltage. Figure 3-5 shows a basic alternator with its single-phase output voltage.

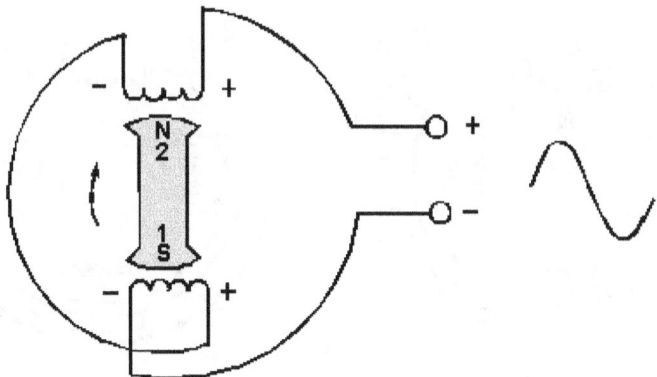

Figure 3-5.—Single-phase alternator.

The definition of phase as you learned it in studying ac circuits may not help too much right here. Remember, "out of phase" meant "out of time."

Now, it may be easier to think of the word *phase* as meaning voltage as in single voltage. The need for a modified definition of phase in this usage will be easier to see as we go along.

Single-phase alternators are found in many applications. They are most often used when the loads being driven are relatively light. The reason for this will be more apparent as we get into multiphase alternators (also called polyphase).

Power that is used in homes, shops, and ships to operate portable tools and small appliances is single-phase power. Single-phase power alternators always generate single-phase power. However, all single-phase power does not come from single-phase alternators. This will sound more reasonable to you as we get into the next subjects.

Q9. What does the term single phase indicate?

Q10. In single-phase alternators, in order for the voltages induced in all the armature windings to add together for a single output, how must the windings be connected?

TWO-PHASE ALTERNATORS

Two phase implies two voltages if we apply our new definition of phase. And, it's that simple. A two-phase alternator is designed to produce two completely separate voltages. Each voltage, by itself, may be considered as a single-phase voltage. Each is generated completely independent of the other. Certain advantages are gained. These and the mechanics of generation will be covered in the following paragraphs.

Generation of Two-Phase Power

Figure 3-6 shows a simplified two-pole, two-phase alternator. Note that the windings of the two phases are physically at right angles (90°) to each other. You would expect the outputs of each phase to be 90° apart, which they are. The graph shows the two phases to be 90° apart, with A leading B. Note that by using our original definition of phase (from previous modules), we could say that A and B are 90° out of phase. There will always be 90° between the phases of a two-phase alternator. This is by design.

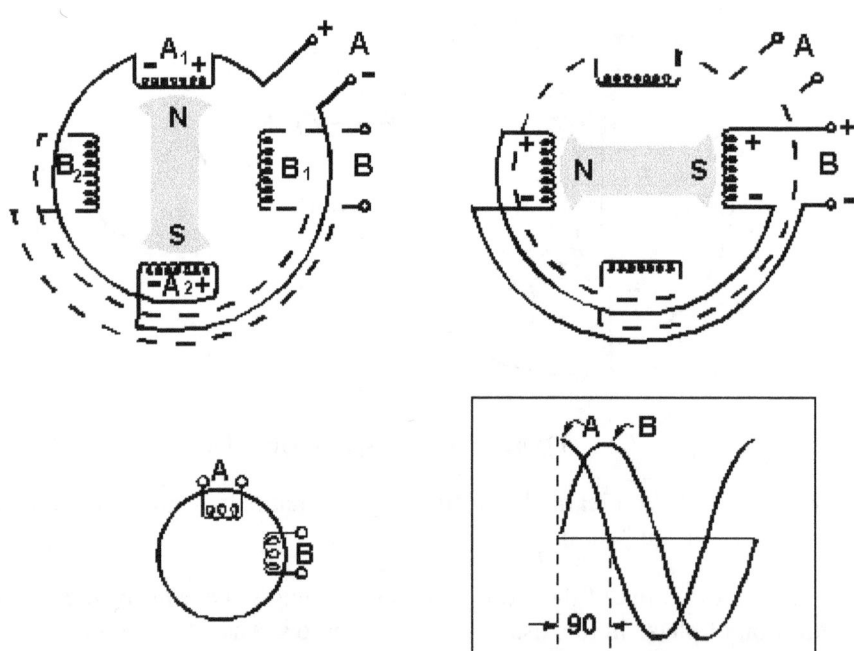

Figure 3-6.—Two-phase alternator.

Now, let's go back and see the similarities and differences between our original (single-phase) alternators and this new one (two-phase). Note that the principles applied are not new. This alternator works the same as the others we have discussed.

The stator in figure 3-6 consists of two single-phase windings completely separated from each other. Each winding is made up of two windings that are connected in series so that their voltages add. The rotor is identical to that used in the single-phase alternator. In the left-hand schematic, the rotor poles are opposite all the windings of phase A. Therefore, the voltage induced in phase A is maximum, and the voltage induced in phase B is zero. As the rotor continues rotating counterclockwise, it moves away from the A windings and approaches the B windings. As a result, the voltage induced in phase A decreases from its maximum value, and the voltage induced in phase B increases from zero. In the right-hand schematic, the rotor poles are opposite the windings of phase B. Now the voltage induced in phase B is maximum, whereas the voltage induced in phase A has dropped to zero. Notice that a 90-degree rotation of the rotor corresponds to one-quarter of a cycle, or 90 electrical degrees. The waveform picture shows the voltages induced in phase A and B for one cycle. The two voltages are 90° out of phase. Notice that the two outputs, A and B, are independent of each other. Each output is a single-phase voltage, just as if the other did not exist.

The obvious advantage, so far, is that we have two separate output voltages. There is some saving in having one set of bearings, one rotor, one housing, and so on, to do the work of two. There is the disadvantage of having twice as many stator coils, which require a larger and more complex stator.

The large schematic in figure 3-7 shows four separate wires brought out from the A and B stator windings. This is the same as in figure 3-6. Notice, however, that the dotted wire now connects one end of B1 to one end of A2. The effect of making this connection is to provide a new output voltage. This sine-wave voltage, C in the picture, is larger than either A or B. It is the result of adding the instantaneous values of phase A and phase B. For this reason it appears exactly half way between A and B. Therefore, C must lag A by 45° and lead B by 45°, as shown in the small vector diagram.

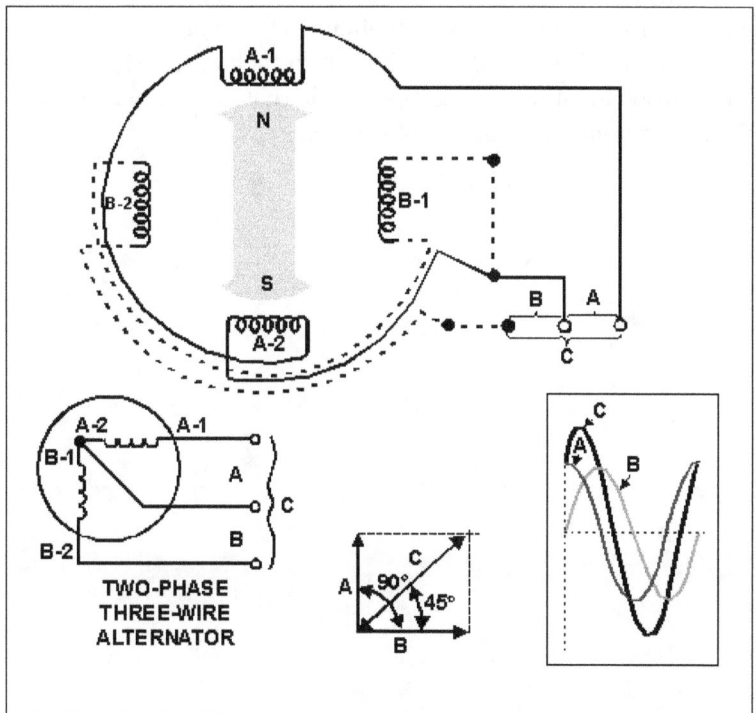

Figure 3-7.—Connections of a two-phase, three-wire alternator output.

Now, look at the smaller schematic diagram in figure 3-7. Only three connections have been brought out from the stator. Electrically, this is the same as the large diagram above it. Instead of being connected at the output terminals, the B1-A2 connection was made internally when the stator was wired. A two-phase alternator connected in this manner is called a two-phase, three-wire alternator.

The three-wire connection makes possible three different load connections: A and B (across each phase), and C (across both phases). The output at C is always 1.414 times the voltage of either phase. These multiple outputs are additional advantages of the two-phase alternator over the single-phase type.

Now, you can understand why single-phase power doesn't always come from single-phase alternators. It can be generated by two-phase alternators as well as other multiphase (polyphase) alternators, as you will soon see.

The two-phase alternator discussed in the preceding paragraphs is seldom seen in actual use. However, the operation of polyphase alternators is more easily explained using two phases than three phases. The three-phase alternator, which will be covered next, is by far the most common of all alternators in use today, both in military and civilian applications.

Q11. What determines the phase relationship between the voltages in a two-phase ac generator?

Q12. How many voltage outputs are available from a two-phase three-wire alternator?

Q13. What is the relationship of the voltage at C in figure 3-7 to the voltages at A and B?

THREE-PHASE ALTERNATOR

The three-phase alternator, as the name implies, has three single-phase windings spaced such that the voltage induced in any one phase is displaced by 120° from the other two. A schematic diagram of a three-phase stator showing all the coils becomes complex, and it is difficult to see what is actually happening. The simplified schematic of figure 3-8, view A, shows all the windings of each phase lumped together as one winding. The rotor is omitted for simplicity. The voltage waveforms generated across each phase are drawn on a graph, phase-displaced 120° from each other. The three-phase alternator as shown in this schematic is made up of three single-phase alternators whose generated voltages are out of phase by 120° . The three phases are independent of each other.

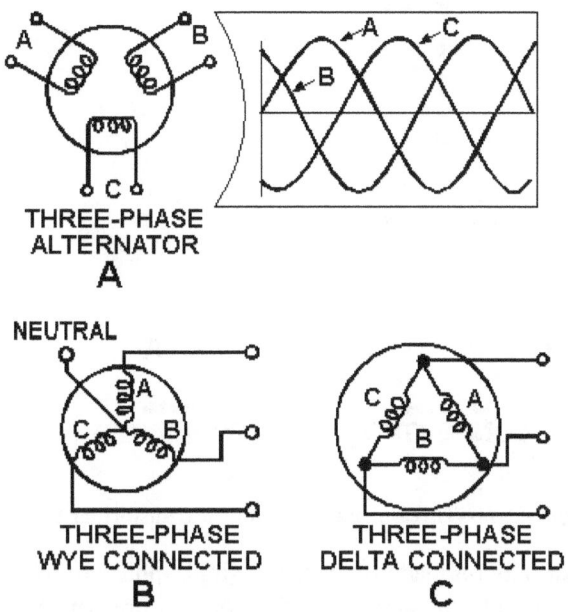

Figure 3-8.—Three-phase alternator connections.

Rather than having six leads coming out of the three-phase alternator, the same leads from each phase may be connected together to form a wye (Y) connection, as shown in figure 3-8, view B. It is called a wye connection because, without the neutral, the windings appear as the letter Y, in this case sideways or upside down.

The neutral connection is brought out to a terminal when a single-phase load must be supplied. Single-phase voltage is available from neutral to A, neutral to B, and neutral to C.

In a three-phase, Y-connected alternator, the total voltage, or line voltage, across any two of the three line leads is the vector sum of the individual phase voltages. Each line voltage is 1.73 times one of the phase voltages. Because the windings form only one path for current flow between phases, the line and phase currents are the same (equal).

A three-phase stator can also be connected so that the phases are connected end-to-end; it is now delta connected (fig. 3-8, view C). (Delta because it looks like the Greek letter delta, Δ.) In the delta connection, line voltages are equal to phase voltages, but each line current is equal to 1.73 times the phase current. Both the wye and the delta connections are used in alternators.

The majority of all alternators in use in the Navy today are three-phase machines. They are much more efficient than either two-phase or single-phase alternators.

Three-Phase Connections

The stator coils of three-phase alternators may be joined together in either wye or delta connections, as shown in figure 3-9. With these connections only three wires come out of the alternator. This allows convenient connection to three-phase motors or power distribution transformers. It is necessary to use three-phase transformers or their electrical equivalent with this type of system.

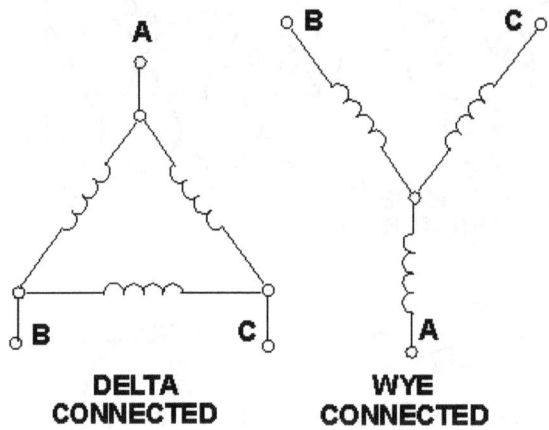

Figure 3-9.—Three-phase alternator or transformer connections.

A three-phase transformer may be made up of three, single-phase transformers connected in delta, wye, or a combination of both. If both the primary and secondary are connected in wye, the transformer is called a wye-wye. If both windings are connected in delta, the transformer is called a delta-delta.

Figure 3-10 shows single-phase transformers connected delta-delta for operation in a three-phase system. You will note that the transformer windings are not angled to illustrate the typical delta (Δ) as has been done with alternator windings. Physically, each transformer in the diagram stands alone. There is no angular relationship between the windings of the individual transformers. However, if you follow the connections, you will see that they form an electrical delta. The primary windings, for example, are connected to each other to form a closed loop. Each of these junctions is fed with a phase voltage from a three-phase alternator. The alternator may be connected either delta or wye depending on load and voltage requirements, and the design of the system.

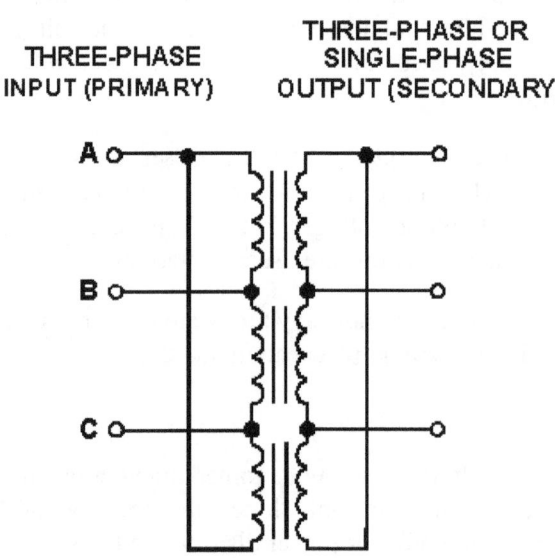

Figure 3-10.—Three single-phase transformers connected delta-delta.

Figure 3-11 shows three single-phase transformers connected wye-wye. Again, note that the transformer windings are not angled. Electrically, a Y is formed by the connections. The lower connections of each winding are shorted together. These form the common point of the wye. The opposite end of each winding is isolated. These ends form the arms of the wye.

THREE-PHASE INPUT (PRIMARY)

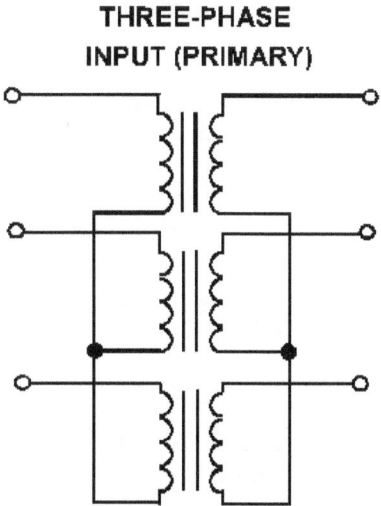

Figure 3-11.—Three single-phase transformers connected wye-wye.

The ac power on most ships is distributed by a three-phase, three-wire, 450-volt system. The single-phase transformers step the voltage down to 117 volts. These transformers are connected delta-delta as in figure 3-10. With a delta-delta configuration, the load may be a three-phase device connected to all phases; or, it may be a single-phase device connected to only one phase.

At this point, it is important to remember that such a distribution system includes everything between the alternator and the load. Because of the many choices that three-phase systems provide, care must be taken to ensure that any change of connections does not provide the load with the wrong voltage or the wrong phase.

Q14. In a three-phase alternator, what is the phase relationship between the individual output voltages?

Q15. What are the two methods of connecting the outputs from a three-phase alternator to the load?

Q16. Ships' generators produce 450-volt, three-phase, ac power; however, most equipment uses 117-volt, single-phase power What transformers and connections are used to convert 450-volt, three-phase power to 117-volt, single-phase power?

FREQUENCY

The output frequency of alternator voltage depends upon the speed of rotation of the rotor and the number of poles. The faster the speed, the higher the frequency. The lower the speed, the lower the frequency. The more poles there are on the rotor, the higher the frequency is for a given speed. When a rotor has rotated through an angle such that two adjacent rotor poles (a north and a south pole) have passed one winding, the voltage induced in that winding will have varied through one complete cycle. For a given frequency, the more pairs of poles there are, the lower the speed of rotation. This principle is illustrated in figure 3-12; a two-pole generator must rotate at four times the speed of an eight-pole generator to produce the same frequency of generated voltage. The frequency of any ac generator in hertz (Hz), which is the number of cycles per second, is related to the number of poles and the speed of rotation, as expressed by the equation

$$F = \frac{NP}{120}$$

where P is the number of poles, N is the speed of rotation in revolutions per minute (rpm), and 120 is a constant to allow for the conversion of minutes to seconds and from poles to pairs of poles. For example, a 2-pole, 3600-rpm alternator has a frequency of 60 Hz; determined as follows:

$$\frac{2 \times 3600}{120} = 60\text{Hz}$$

A 4-pole, 1800-rpm generator also has a frequency of 60 Hz. A 6-pole, 500-rpm generator has a frequency of

$$\frac{6 \times 500}{120} = 25\text{Hz}$$

A 12-pole, 4000-rpm generator has a frequency of

$$\frac{12 \times 4000}{120} = 400\text{Hz}$$

Q17. *What two factors determine the frequency of the output voltage of an alternator?*

Q18. *What is the frequency of the output voltage of an alternator with four poles that is rotated at 3600 rpm?*

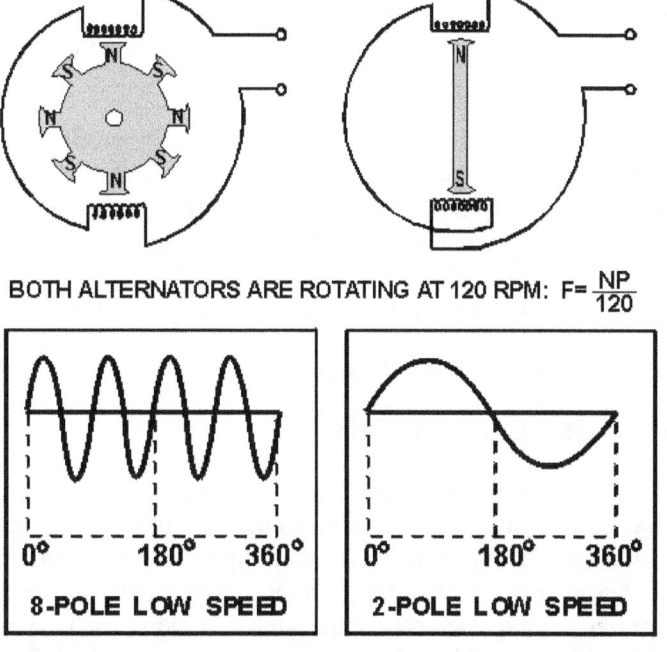

$$\text{BOTH ALTERNATORS ARE ROTATING AT 120 RPM: } F = \frac{NP}{120}$$

Figure 3-12.—Frequency regulation.

VOLTAGE REGULATION

As we have seen before, when the load on a generator is changed, the terminal voltage varies. The amount of variation depends on the design of the generator.

The voltage regulation of an alternator is the change of voltage from full load to no load, expressed as a percentage of full-load volts, when the speed and dc field current are held constant.

$$\frac{E_{nL} - E_{fL}}{E_{fL}} \times 100 = \text{Percent of regulation}$$

Assume the no-load voltage of an alternator is 250 volts and the full-load voltage is 220 volts. The percent of regulation is

$$\frac{250 - 220}{220} \times 100 = 13.6\%$$

Remember, the lower the percent of regulation, the better it is in most applications.

Q19. The variation in output voltage as the load changes is referred to as what? How is it expressed?

PRINCIPLES OF AC VOLTAGE CONTROL

In an alternator, an alternating voltage is induced in the armature windings when magnetic fields of alternating polarity are passed across these windings. The amount of voltage induced in the windings depends mainly on three things: (1) the number of conductors in series per winding, (2) the speed (alternator rpm) at which the magnetic field cuts the winding, and (3) the strength of the magnetic field. Any of these three factors could be used to control the amount of voltage induced in the alternator windings.

The number of windings, of course, is fixed when the alternator is manufactured. Also, if the output frequency is required to be of a constant value, then the speed of the rotating field must be held constant. This prevents the use of the alternator rpm as a means of controlling the voltage output. Thus, the only practical method for obtaining voltage control is to control the strength of the rotating magnetic field. The strength of this electromagnetic field may be varied by changing the amount of current flowing through the field coil. This is accomplished by varying the amount of voltage applied across the field cod.

Q20. How is output voltage controlled in practical alternators?

PARALLEL OPERATION OF ALTERNATORS

Alternators are connected in parallel to (1) increase the output capacity of a system beyond that of a single unit, (2) serve as additional reserve power for expected demands, or (3) permit shutting down one machine and cutting in a standby machine without interrupting power distribution. When alternators are of sufficient size, and are operating at different frequencies and terminal voltages, severe damage may result if they are suddenly connected to each other through a common bus. To avoid this, the machines must be synchronized as closely as possible before connecting them together. This may be accomplished by connecting one generator to the bus (referred to as bus generator), and then synchronizing the other (incoming generator) to it before closing the incoming generator's main power contactor. The generators are synchronized when the following conditions are set:

1. Equal terminal voltages. This is obtained by adjustment of the incoming generator's field strength.

2. Equal frequency. This is obtained by adjustment of the incoming generator's prime-mover speed.

3. Phase voltages in proper phase relation. The procedure for synchronizing generators is not discussed in this chapter. At this point, it is enough for you to know that the above must be accomplished to prevent damage to the machines.

Q21. What generator characteristics must be considered when alternators are synchronized for parallel operation?

SUMMARY

This chapter has presented an introduction to the subject of alternators. You have studied the characteristics and applications of different types. The following information provides a summary of the chapter for your review.

MAGNETIC INDUCTION is the process of inducing an emf in a coil whenever the coil is placed in a magnetic field and motion exists between the coil and the magnetic lines of flux. This is true if either the coil or the magnetic field moves, as long as the coil is caused to cut across magnetic flux lines.

The **ROTATING ARMATURE-ALTERNATOR** is essentially a loop rotating through a stationary magnetic fealties cutting action of the loop through the magnetic field generates ac in the loop. This ac is removed from the loop by means of slip rings and applied to an external load.

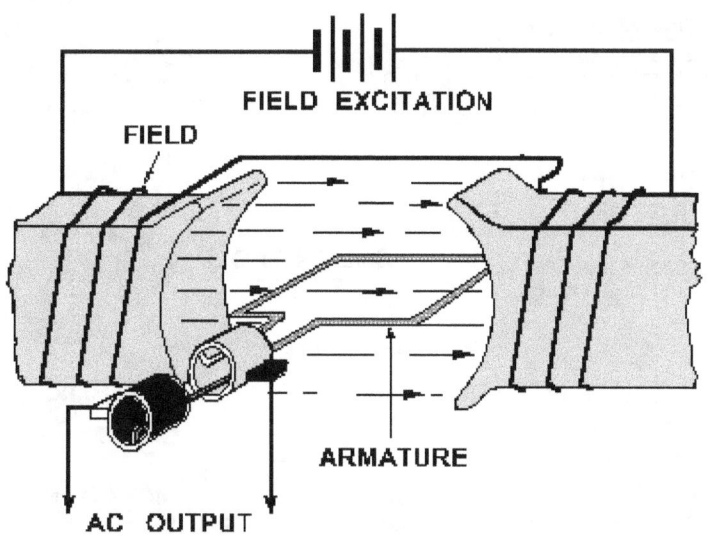

The **ROTATING-FIELD ALTERNATOR** has a stationary armature and a rotating field. High voltages can be generated in the armature and applied to the load directly, without the need of slip rings and brushes. The low dc voltage is applied to the rotor field by means of slip rings, but this does not introduce any insulation problems.

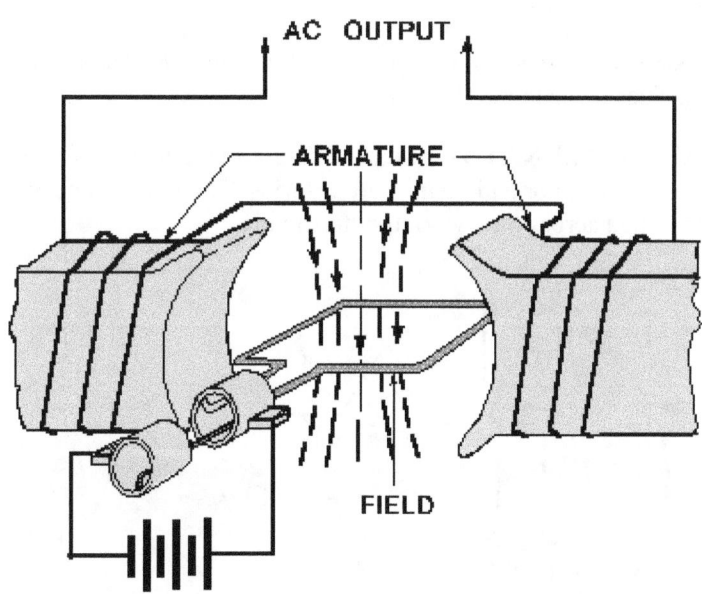

ROTOR CONSTRUCTION in alternators may be either of two types. The salient-pole rotor is used in slower speed alternators. The turbine driven-type is wound in a manner to allow high-speed use without flying apart.

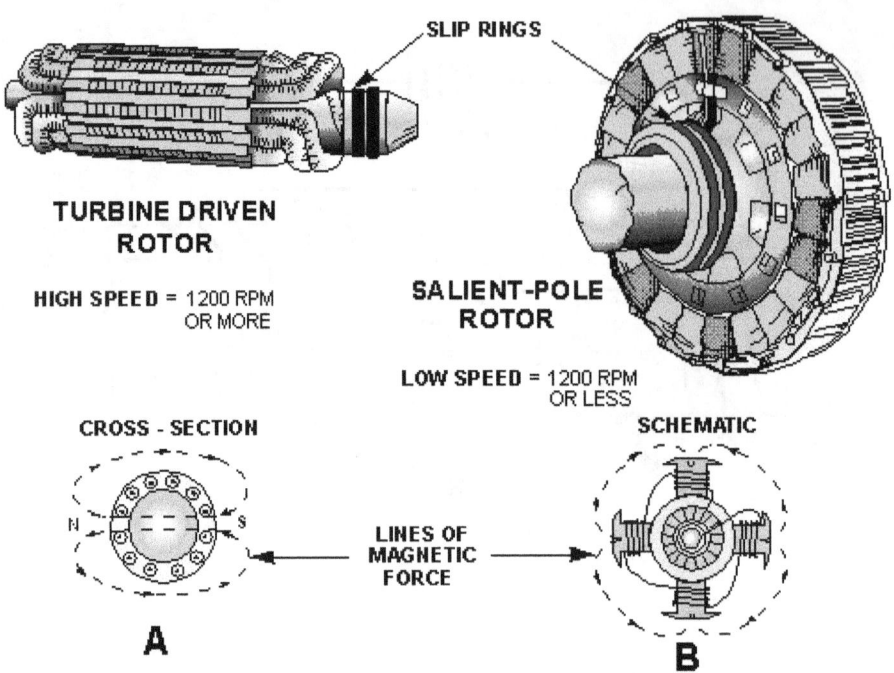

GENERATOR RATINGS are dependent on the amount of current they are capable of providing at full output voltage; this rating is expressed as the product of the voltage times the current. A 10-volt alternator capable of supplying 10 amperes of current would be rated at 100 volt-amperes. Larger alternators are rated in kilovolt-amperes.

EXCITER GENERATORS are small dc generators built into alternators to provide excitation current to field windings. These dc generators are called exciters.

The **SINGLE-PHASE ALTERNATOR** has an armature that consists of a number of windings placed symmetrically around the stator and connected in series. The voltages generated in each winding add to produce the total voltage across the two output terminals.

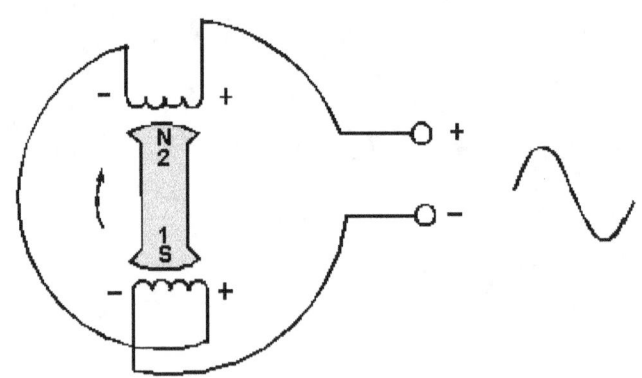

A **TWO-PHASE ALTERNATOR** consists of two phases whose windings are so placed around the stator that the voltages generated in them are 90° out of phase.

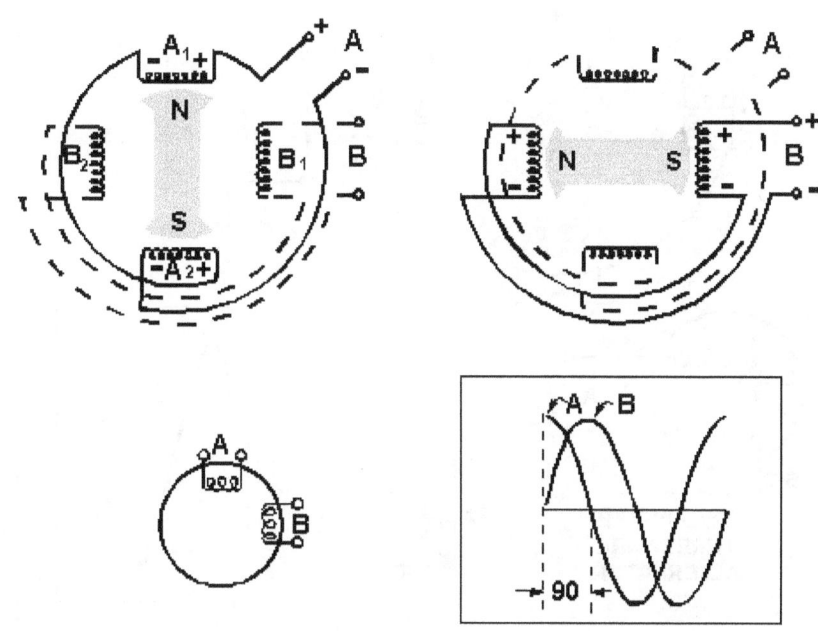

TWO-PHASE ALTERNATOR CONNECTIONS may be modified so that the output of a two-phase alternator is in a three-wire manner, which actually provides three outputs, two induced phase voltages, plus a vectorial sum voltage.

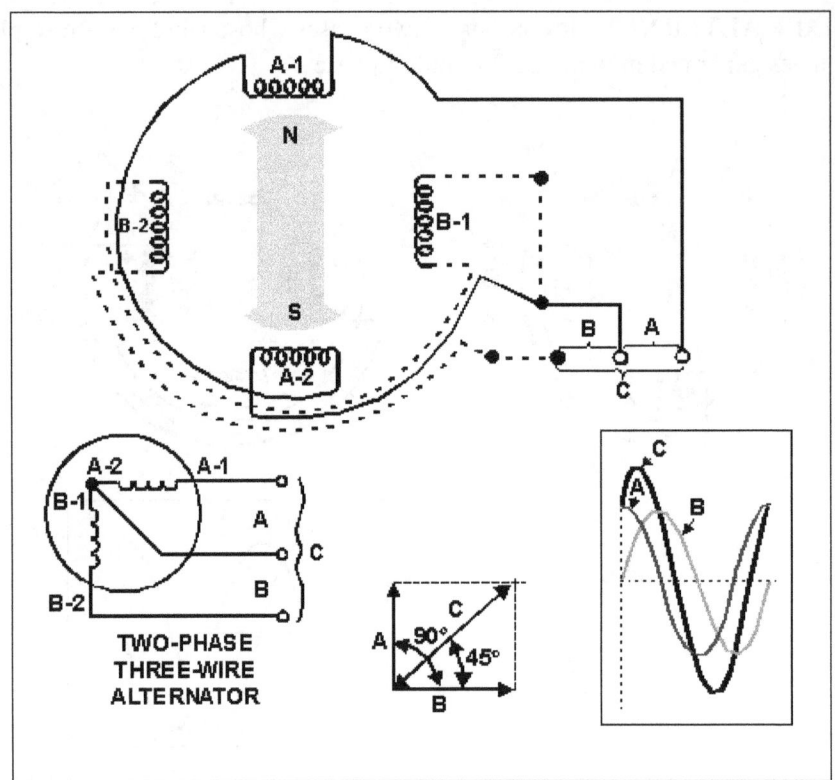

TWO-PHASE
THREE-WIRE
ALTERNATOR

In **THREE-PHASE ALTERNATORS** the windings have voltages generated in them which are 120° out of phase. Three-phase alternators are most often used to generate ac power.

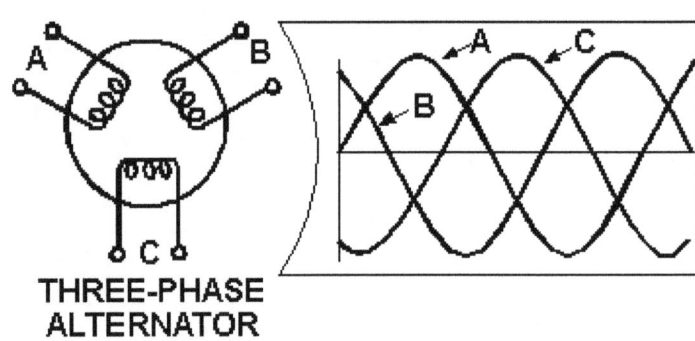

THREE-PHASE
ALTERNATOR

THREE-PHASE ALTERNATOR CONNECTIONS may be delta or wye connections depending on the application. The ac power aboard ship is usually taken from the ship's generators through delta connections, for the convenience of step-down transformers.

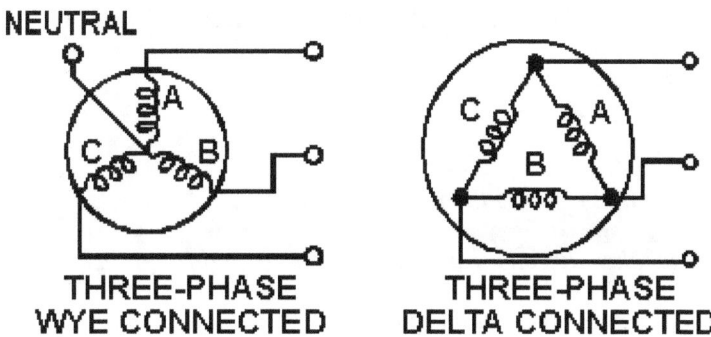

THREE-PHASE
WYE CONNECTED

THREE-PHASE
DELTA CONNECTED

ALTERNATOR FREQUENCY depends upon the speed of rotation and the number of pairs of rotor poles.

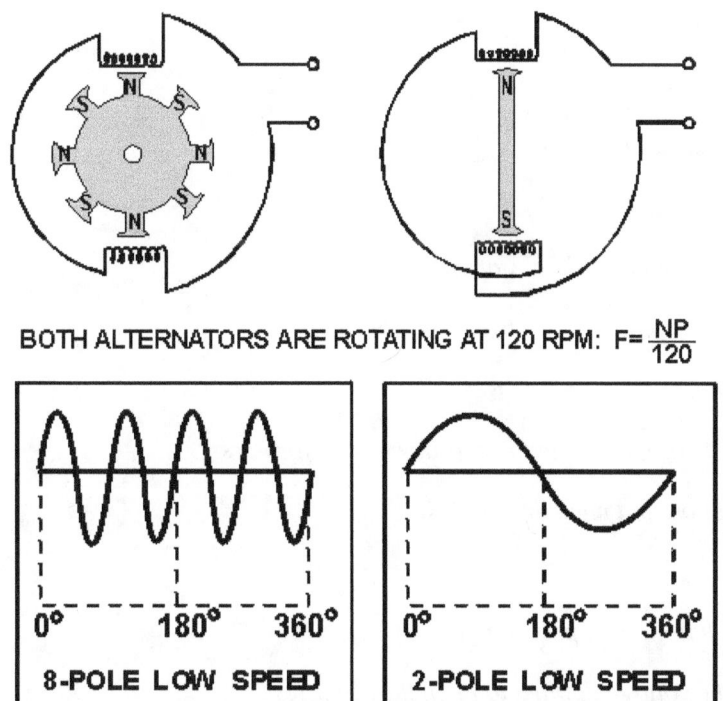

BOTH ALTERNATORS ARE ROTATING AT 120 RPM: $F = \dfrac{NP}{120}$

8-POLE LOW SPEED

2-POLE LOW SPEED

VOLTAGE REGULATION is the change in output voltage of an alternator under varying load conditions.

VOLTAGE CONTROL in alternators is accomplished by varying the current in the field windings, much as in dc generators.

ANSWERS TO QUESTIONS Q1. THROUGH Q21.

A1.   *A conductor and a magnetic field.*

A2.   *Armature.*

A3.   *Rotating armature and rotating field.*

A4.   *Output voltage is taken directly from the armature (not through brushes or slip rings).*

A5.   *To provide dc current for the rotating field.*

A6.   *Kilovolt-amperes (volt amperes).*

A7.   *Steam turbine.*

A8.   *Internal combustion engines, water force and electric motors.*

A9.   *One voltage (one output).*

A10.   *In series.*

A11.   *Placement of armature coils.*

A12.   *Three.*

A13.   *C is 1.414 times greater than A or B.*

A14.   *Each phase is displaced 120° from the other two.*

A15.   *Wye and Delta.*

A16.   *Three single-phase, delta-delta, step-down transformers.*

A17.   *Speed of rotation and number of poles.*

A18.   *120 Hz.*

A19.   *Voltage regulation. As a percentage.*

A20.   *By varying the voltage applied to the field windings.*

A21.   *Output voltage, frequency, and phase relationships.*

CHAPTER 4

ALTERNATING CURRENT MOTORS

LEARNING OBJECTIVES

Upon completion of this chapter you will be able to:

1. List three basic types of ac motors and describe the characteristics of each type.

2. Describe the characteristics of a series motor that enable it to be used as a universal motor.

3. Explain the relationships of the individual phases of multiphase voltages as they produce rotating magnetic fields in ac motors.

4. Describe the placement of stator windings in two-phase, ac motors using rotating fields.

5. List the similarities and differences between the stator windings of two-phase and three-phase ac motors.

6. State the primary application of synchronous motors, and explain the characteristics that make them suitable for that application.

7. Describe the features that make the ac induction motor the most widely used of electric motors.

8. Describe the difference between the rotating field of multiphase motors and the "apparent" rotating field of single-phase motors.

9. Explain the operation of split-phase windings in single-phase ac induction motors.

10. Describe the effects of shaded poles in single-phase, ac induction motors.

INTRODUCTION

Most of the power-generating systems, ashore and afloat, produce ac. For this reason a majority of the motors used throughout the Navy are designed to operate on ac. There are other advantages in the use of ac motors besides the wide availability of ac power. In general, ac motors cost less than dc motors. Some types of ac motors do not use brushes and commutators. This eliminates many problems of maintenance and wear. It also eliminates the problem of dangerous sparking.

An ac motor is particularly well suited for constant-speed applications. This is because its speed is determined by the frequency of the ac voltage applied to the motor terminals.

The dc motor is better suited than an ac motor for some uses, such as those that require variable-speeds. An ac motor can also be made with variable speed characteristics but only within certain limits.

Industry builds ac motors in different sizes, shapes, and ratings for many different types of jobs. These motors are designed for use with either polyphase or single-phase power systems. It is not possible here to cover all aspects of the subject of ac motors. Only the principles of the most commonly used types are dealt with in this chapter.

In this chapter, ac motors will be divided into (1) series, (2) synchronous, and (3) induction motors. Single-phase and polyphase motors will be discussed. Synchronous motors, for purposes of this chapter, may be considered as polyphase motors, of constant speed, whose rotors are energized with dc voltage. Induction motors, single-phase or polyphase, whose rotors are energized by induction, are the most commonly used ac motor. The series ac motor, in a sense, is a familiar type of motor. It is very similar to the dc motor that was covered in chapter 2 and will serve as a bridge between the old and the new.

Q1. What are the three basic types of ac motors?

SERIES AC MOTOR

A series ac motor is the same electrically as a dc series motor. Refer to figure 4-1 and use the left-hand rule for the polarity of coils. You can see that the instantaneous magnetic polarities of the armature and field oppose each other, and motor action results. Now, reverse the current by reversing the polarity of the input. Note that the field magnetic polarity still opposes the armature magnetic polarity. This is because the reversal effects both the armature and the field. The ac input causes these reversals to take place continuously.

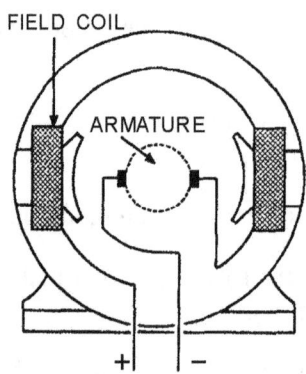

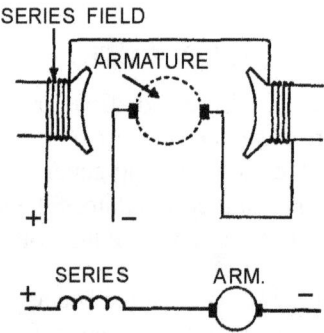

Figure 4-1.—Series ac motor.

The construction of the ac series motor differs slightly from the dc series motor. Special metals, laminations, and windings are used. They reduce losses caused by eddy currents, hysteresis, and high reactance. Dc power can be used to drive an ac series motor efficiently, but the opposite is not true.

The characteristics of a series ac motor are similar to those of a series dc motor. It is a varying-speed machine. It has low speeds for large loads and high speeds for light loads. The starting torque is very high. Series motors are used for driving fans, electric drills, and other small appliances. Since the series ac motor has the same general characteristics as the series dc motor, a series motor has been designed that can operate both on ac and dc. This ac/dc motor is called a universal motor. It finds wide use in small electric appliances. Universal motors operate at lower efficiency than either the ac or dc series motor. They are built in small sizes only. Universal motors do not operate on polyphase ac power.

Q2. Series motors are generally used to operate what type of equipment?

Q3. Why are series motors sometimes called universal motors?

ROTATING MAGNETIC FIELDS

The principle of rotating magnetic fields is the key to the operation of most ac motors. Both synchronous and induction types of motors rely on rotating magnetic fields in their stators to cause their rotors to turn.

The idea is simple. A magnetic field in a stator can be made to rotate electrically, around and around. Another magnetic field in the rotor can be made to chase it by being attracted and repelled by the stator field. Because the rotor is free to turn, it follows the rotating magnetic field in the stator. Let's see how it is done.

Rotating magnetic fields may be set up in two-phase or three-phase machines. To establish a rotating magnetic field in a motor stator, the number of pole pairs must be the same as (or a multiple of) the number of phases in the applied voltage. The poles must then be displaced from each other by an angle equal to the phase angle between the individual phases of the applied voltage.

Q4. What determines the number of field poles required to establish a rotating magnetic field in a multiphase motor stator?

TWO-PHASE ROTATING MAGNETIC FIELD

A rotating magnetic field is probably most easily seen in a two-phase stator. The stator of a two-phase induction motor is made up of two windings (or a multiple of two). They are placed at right angles to each other around the stator. The simplified drawing in figure 4-2 illustrates a two-phase stator.

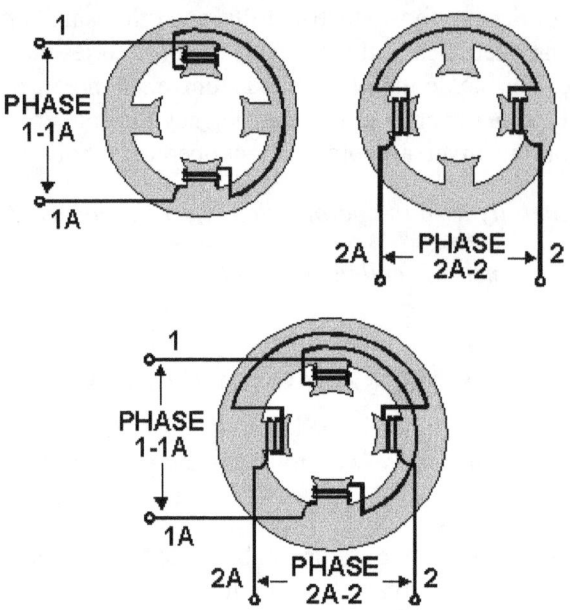

Figure 4-2.—Two-phase motor stator.

If the voltages applied to phases 1-1A and 2-2A are 90° out of phase, the currents that flow in the phases are displaced from each other by 90° . Since the magnetic fields generated in the coils are in phase with their respective currents, the magnetic fields are also 90° out of phase with each other. These two out-of-phase magnetic fields, whose coil axes are at right angles to each other, add together at every instant during their cycle. They produce a resultant field that rotates one revolution for each cycle of ac.

To analyze the rotating magnetic field in a two-phase stator, refer to figure 4-3. The arrow represents the rotor. For each point set up on the voltage chart, consider that current flows in a direction that will cause the magnetic polarity indicated at each pole piece. Note that from one point to the next, the polarities are rotating from one pole to the next in a clockwise manner. One complete cycle of input voltage produces a 360-degree rotation of the pole polarities. Let's see how this result is obtained.

4-4

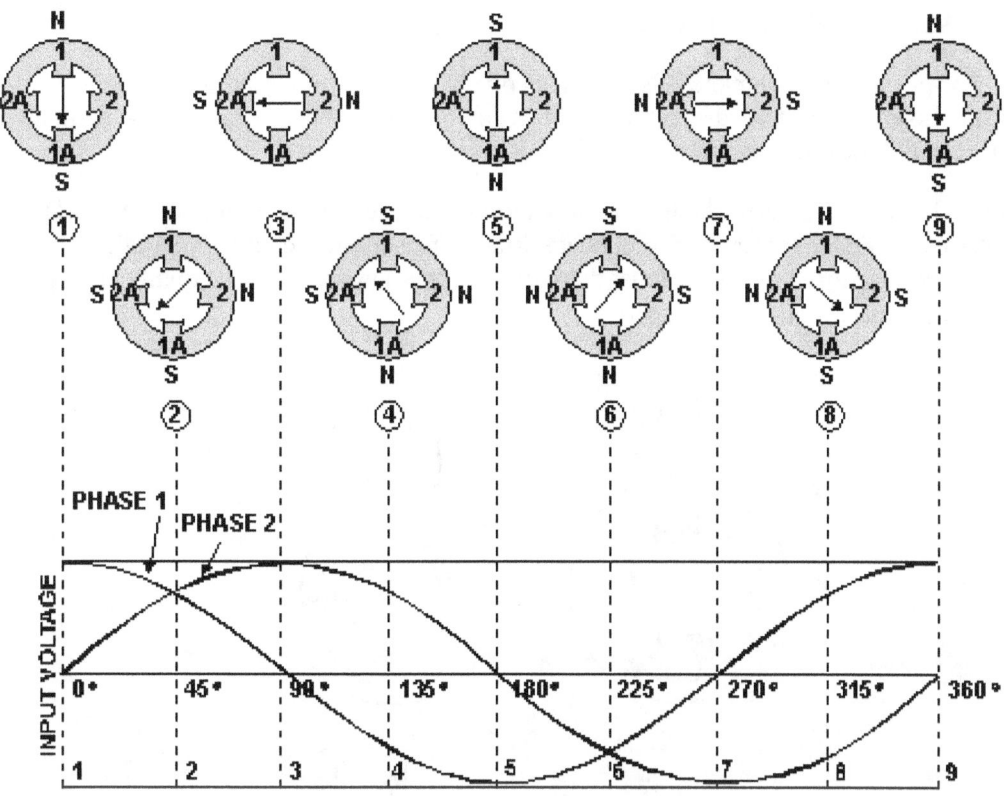

Figure 4-3.—Two-phase rotating field.

The waveforms in figure 4-3 are of the two input phases, displaced 90° because of the way they were generated in a two-phase alternator. The waveforms are numbered to match their associated phase. Although not shown in this figure, the windings for the poles 1-1A and 2-2A would be as shown in the previous figure. At position 1, the current flow and magnetic field in winding 1-1A is at maximum (because the phase voltage is maximum). The current flow and magnetic field in winding 2-2A is zero (because the phase voltage is zero). The resultant magnetic field is therefore in the direction of the 1-1A axis. At the 45-degree point (position 2), the resultant magnetic field lies midway between windings 1-1A and 2-2A. The coil currents and magnetic fields are equal in strength. At 90° (position 3), the magnetic field in winding 1-1A is zero. The magnetic field in winding 2-2A is at maximum. Now the resultant magnetic field lies along the axis of the 2-2A winding as shown. The resultant magnetic field has rotated clockwise through 90° to get from position 1 to position 3. When the two-phase voltages have completed one full cycle (position 9), the resultant magnetic field has rotated through 360° . Thus, by placing two windings at right angles to each other and exciting these windings with voltages 90° out of phase, a rotating magnetic field results.

Two-phase motors are rarely used except in special-purpose equipment. They are discussed here to aid in understanding rotating fields. You will, however, encounter many single-phase and three-phase motors.

Q5.   What is the angular displacement between field poles in a two-phase motor stator?

THREE-PHASE ROTATING FIELDS

The three-phase induction motor also operates on the principle of a rotating magnetic field. The following discussion shows how the stator windings can be connected to a three-phase ac input and have a resultant magnetic field that rotates.

Figure 4-4, views A-C show the individual windings for each phase. Figure 4-4, view D, shows how the three phases are tied together in a Y-connected stator. The dot in each diagram indicates the common point of the Y-connection. You can see that the individual phase windings are equally spaced around the stator. This places the windings 120° apart.

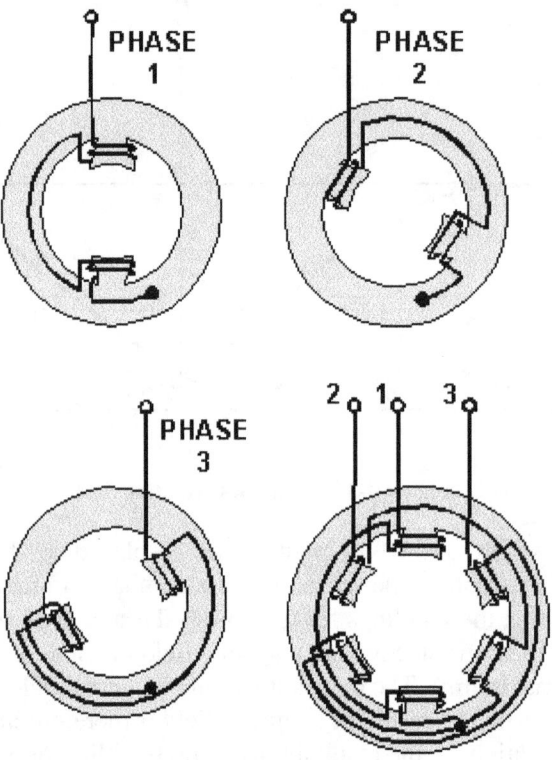

Figure 4-4.—Three-phase, Y-connected stator.

The three-phase input voltage to the stator of figure 4-4 is shown in the graph of figure 4-5. Use the left-hand rule for determining the electromagnetic polarity of the poles at any given instant. In applying the rule to the coils in figure 4-4, consider that current flows toward the terminal numbers for positive voltages, and away from the terminal numbers for negative voltages.

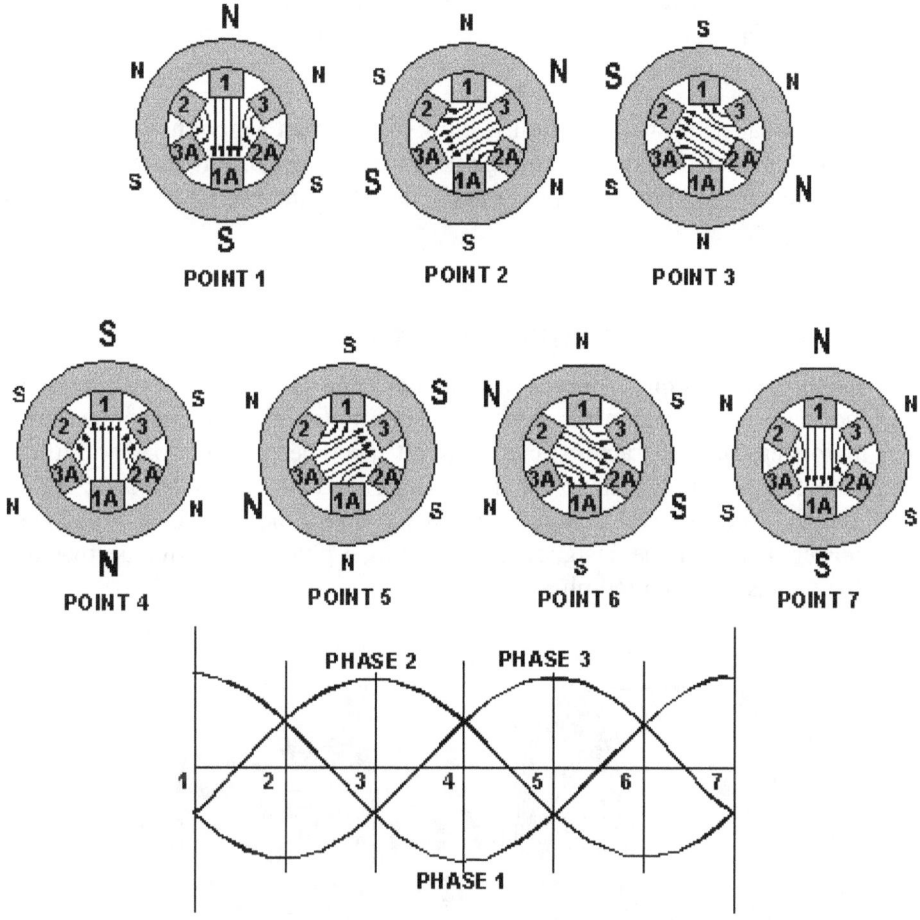

Figure 4-5.—Three-phase rotating-field polarities and input voltages.

The results of this analysis are shown for voltage points 1 through 7 in figure 4-5. At point 1, the magnetic field in coils 1-1A is maximum with polarities as shown. At the same time, negative voltages are being felt in the 2-2A and 3-3A windings. These create weaker magnetic fields, which tend to aid the 1-1A field. At point 2, maximum negative voltage is being felt in the 3-3A windings. This creates a strong magnetic field which, in turn, is aided by the weaker fields in 1-1A and 2-2A. As each point on the voltage graph is analyzed, it can be seen that the resultant magnetic field is rotating in a clockwise direction. When the three-phase voltage completes one full cycle (point 7), the magnetic field has rotated through 360° .

Q6.   What is the major difference between a two-phase and a three-phase stator?

ROTOR BEHAVIOR IN A ROTATING FIELD

For purposes of explaining rotor movement, let's assume that we can place a bar magnet in the center of the stator diagrams of figure 4-5. We'll mount this magnet so that it is free to rotate in this area. Let's also assume that the bar magnet is aligned so that at point 1 its south pole is opposite the large N of the stator field.

You can see that this alignment is natural. Unlike poles attract, and the two fields are aligned so that they are attracting. Now, go from point 1 through point 7. As before, the stator field rotates clockwise. The bar magnet, free to move, will follow the stator field, because the attraction between the two fields continues to exist. A shaft running through the pivot point of the bar magnet would rotate at the same speed as the rotating field. This speed is known as synchronous speed. The shaft represents the shaft of an operating motor to which the load is attached.

Remember, this explanation is an oversimplification. It is meant to show how a rotating field can cause mechanical rotation of a shaft. Such an arrangement would work, but it is not used. There are limitations to a permanent magnet rotor. Practical motors use other methods, as we shall see in the next paragraphs.

SYNCHRONOUS MOTORS

The construction of the synchronous motors is essentially the same as the construction of the salient-pole alternator. In fact, such an alternator may be run as an ac motor. It is similar to the drawing in figure 4-6. Synchronous motors have the characteristic of constant speed between no load and full load. They are capable of correcting the low power factor of an inductive load when they are operated under certain conditions. They are often used to drive dc generators. Synchronous motors are designed in sizes up to thousands of horsepower. They may be designed as either single-phase or multiphase machines. The discussion that follows is based on a three-phase design.

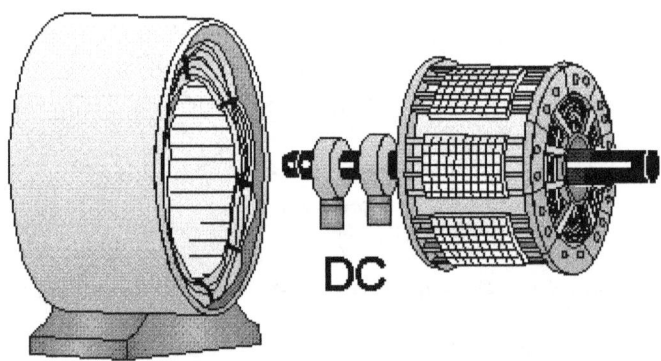

Figure 4-6.—Revolving-field synchronous motor.

To understand how the synchronous motor works, assume that the application of three-phase ac power to the stator causes a rotating magnetic field to be set up around the rotor. The rotor is energized with dc (it acts like a bar magnet). The strong rotating magnetic field attracts the strong rotor field activated by the dc. This results in a strong turning force on the rotor shaft. The rotor is therefore able to turn a load as it rotates in step with the rotating magnetic field.

It works this way once it's started. However, one of the disadvantages of a synchronous motor is that it cannot be started from a standstill by applying three-phase ac power to the stator. When ac is applied to the stator, a high-speed rotating magnetic field appears immediately. This rotating field rushes past the rotor poles so quickly that the rotor does not have a chance to get started. In effect, the rotor is repelled first in one direction and then the other. A synchronous motor in its purest form has no starting torque. It has torque only when it is running at synchronous speed.

A squirrel-cage type of winding is added to the rotor of a synchronous motor to cause it to start. The squirrel cage is shown as the outer part of the rotor in figure 4-7. It is so named because it is shaped and looks something like a turnable squirrel cage. Simply, the windings are heavy copper bars shorted together by copper rings. A low voltage is induced in these shorted windings by the rotating three-phase stator field. Because of the short circuit, a relatively large current flows in the squirrel cage. This causes a magnetic field that interacts with the rotating field of the stator. Because of the interaction, the rotor begins to turn, following the stator field; the motor starts. We will run into squirrel cages again in other applications, where they will be covered in more detail.

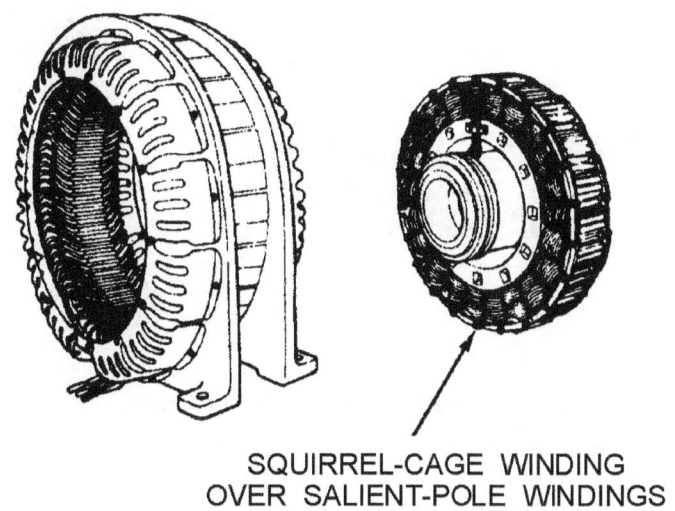

SQUIRREL-CAGE WINDING
OVER SALIENT-POLE WINDINGS

Figure 4-7.—Self-starting synchronous ac motor.

To start a practical synchronous motor, the stator is energized, but the dc supply to the rotor field is not energized. The squirrel-cage windings bring the rotor to near synchronous speed. At that point, the dc field is energized. This locks the rotor in step with the rotating stator field. Full torque is developed, and the load is driven. A mechanical switching device that operates on centrifugal force is often used to apply dc to the rotor as synchronous speed is reached.

The practical synchronous motor has the disadvantage of requiring a dc exciter voltage for the rotor. This voltage may be obtained either externally or internally, depending on the design of the motor.

Q7. What requirement is the synchronous motor specifically designed to meet?

INDUCTION MOTORS

The induction motor is the most commonly used type of ac motor. Its simple, rugged construction costs relatively little to manufacture. The induction motor has a rotor that is not connected to an external source of voltage. The induction motor derives its name from the fact that ac voltages are induced in the rotor circuit by the rotating magnetic field of the stator. In many ways, induction in this motor is similar to the induction between the primary and secondary windings of a transformer.

Large motors and permanently mounted motors that drive loads at fairly constant speed are often induction motors. Examples are found in washing machines, refrigerator compressors, bench grinders, and table saws.

The stator construction of the three-phase induction motor and the three-phase synchronous motor are almost identical. However, their rotors are completely different (see fig. 4-8). The induction rotor is made of a laminated cylinder with slots in its surface. The windings in these slots are one of two types (shown in fig. 4-9). The most common is the squirrel-cage winding. This entire winding is made up of heavy copper bars connected together at each end by a metal ring made of copper or brass. No insulation is required between the core and the bars. This is because of the very low voltages generated in the rotor bars. The other type of winding contains actual coils placed in the rotor slots. The rotor is then called a wound rotor.

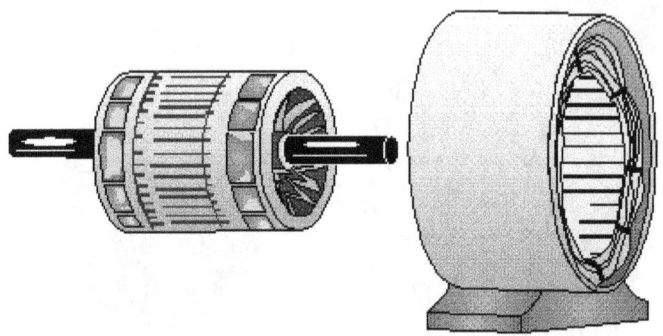

Figure 4-8.—Induction motor.

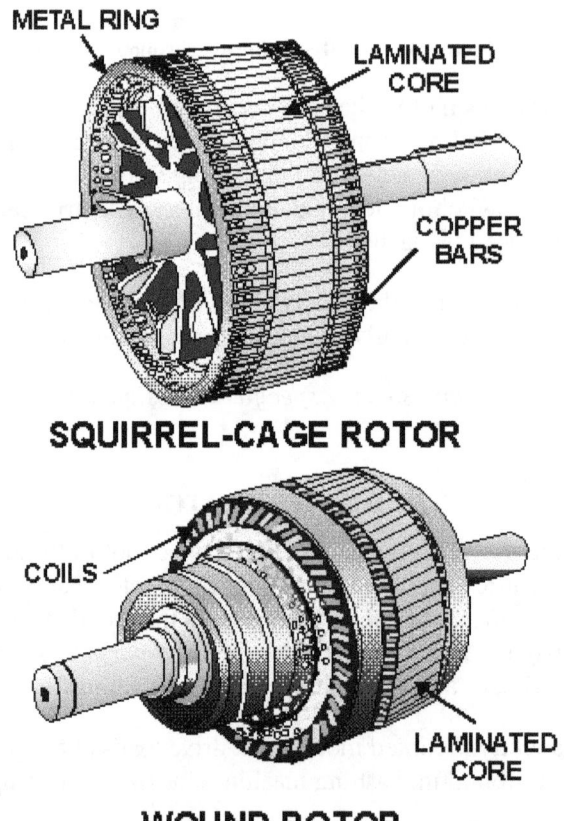

Figure 4-9.—Types of ac induction motor rotors.

Regardless of the type of rotor used, the basic principle is the same. The rotating magnetic field generated in the stator induces a magnetic field in the rotor. The two fields interact and cause the rotor to turn. To obtain maximum interaction between the fields, the air gap between the rotor and stator is very small.

As you know from Lenz's law, any induced emf tries to oppose the changing field that induces it. In the case of an induction motor, the changing field is the motion of the resultant stator field. A force is exerted on the rotor by the induced emf and the resultant magnetic field. This force tends to cancel the relative motion between the rotor and the stator field. The rotor, as a result, moves in the same direction as the rotating stator field.

It is impossible for the rotor of an induction motor to turn at the same speed as the rotating magnetic field. If the speeds were the same, there would be no relative motion between the stator and rotor fields; without relative motion there would be no induced voltage in the rotor. In order for relative motion to exist between the two, the rotor must rotate at a speed slower than that of the rotating magnetic field. The. difference between the speed of the rotating stator field and the rotor speed is called slip. The smaller the slip, the closer the rotor speed approaches the stator field speed.

The speed of the rotor depends upon the torque requirements of the load. The bigger the load, the stronger the turning force needed to rotate the rotor. The turning force can increase only if the rotor-induced emf increases. This emf can increase only if the magnetic field cuts through the rotor at a faster rate. To increase the relative speed between the field and rotor, the rotor must slow down. Therefore, for heavier loads the induction motor turns slower than for lighter loads. You can see from the previous statement that slip is directly proportional to the load on the motor. Actually only a slight change in speed is necessary to produce the usual current changes required for normal changes in load. This is because the rotor windings have such a low resistance. As a result, induction motors are called constant-speed motors.

Q8.   Why is the ac induction motor used more often than other types?

Q9.   The speed of the rotor is always somewhat less than the speed of the rotating field. What is the difference called?

Q10.   What determines the amount of slip in an induction motor?

SINGLE-PHASE INDUCTION MOTORS

There are probably more single-phase ac induction motors in use today than the total of all the other types put together.

It is logical that the least expensive, lowest maintenance type of ac motor should be used most often. The single-phase ac induction motor fits that description.

Unlike polyphase induction motors, the stator field in the single-phase motor does not rotate. Instead it simply alternates polarity between poles as the ac voltage changes polarity.

Voltage is induced in the rotor as a result of magnetic induction, and a magnetic field is produced around the rotor. This field will always be in opposition to the stator field (Lenz's law applies). The interaction between the rotor and stator fields will not produce rotation, however. The interaction is shown by the double-ended arrow in figure 4-10, view A. Because this force is across the rotor and through the pole pieces, there is no rotary motion, just a push and/or pull along this line.

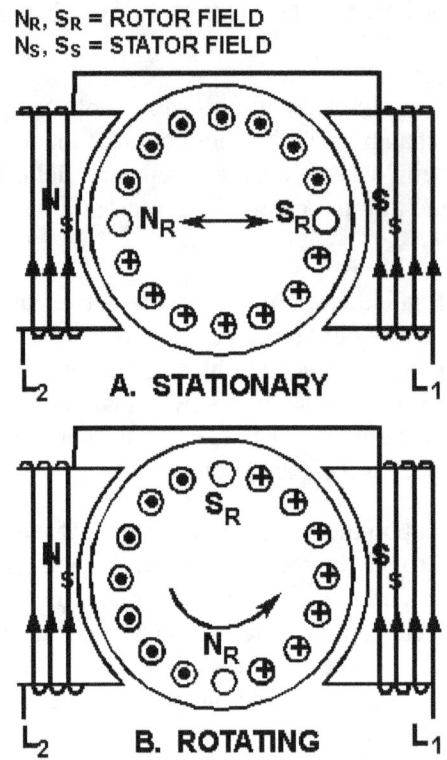

N_R, S_R = ROTOR FIELD
N_S, S_S = STATOR FIELD

A. STATIONARY

B. ROTATING

Figure 4-10.—Rotor currents in a single-phase ac induction motor.

Now, if the rotor is rotated by some outside force (a twist of your hand, or something), the push-pull along the line in figure 4-10, view A, is disturbed. Look at the fields as shown in figure 4-10, view B. At this instant the south pole on the rotor is being attracted by the left-hand pole. The north rotor pole is being attracted to the right-hand pole. All of this is a result of the rotor being rotated 90° by the outside force. The pull that now exists between the two fields becomes a rotary force, turning the rotor toward magnetic correspondence with the stator. Because the two fields continuously alternate, they will never actually line up, and the rotor will continue to turn once started. It remains for us to learn practical methods of getting the rotor to start.

There are several types of single-phase induction motors in use today. Basically they are identical except for the means of starting. In this chapter we will discuss the split-phase and shaded-pole motors; so named because of the methods employed to get them started. Once they are up to operating speed, all single-phase induction motors operate the same.

Q11. What type of ac motor is most widely used?

Split-Phase Induction Motors

One type of induction motor, which incorporates a starting device, is called a split-phase induction motor. Split-phase motors are designed to use inductance, capacitance, or resistance to develop a starting torque. The principles are those that you learned in your study of alternating current.

CAPACITOR-START.—The first type of split-phase induction motor that will be covered is the capacitor-start type. Figure 4-11 shows a simplified schematic of a typical capacitor-start motor. The stator consists of the main winding and a starting winding (auxiliary). The starting winding is connected in parallel with the main winding and is placed physically at right angles to it. A 90-degree electrical phase difference between the two windings is obtained by connecting the auxiliary winding in series with a capacitor and starting switch. When the motor is first energized, the starting switch is closed. This places the capacitor in series with the auxiliary winding. The capacitor is of such value that the auxiliary circuit is effectively a resistive-capacitive circuit (referred to as capacitive reactance and expressed as X_C). In this circuit the current leads the line voltage by about 45° (because X_C about equals R). The main winding has enough resistance-inductance (referred to as inductive reactance and expressed as X_L) to cause the current to lag the line voltage by about 45° (because X_L about equals R). The currents in each winding are therefore 90° out of phase - so are the magnetic fields that are generated. The effect is that the two windings act like a two-phase stator and produce the rotating field required to start the motor.

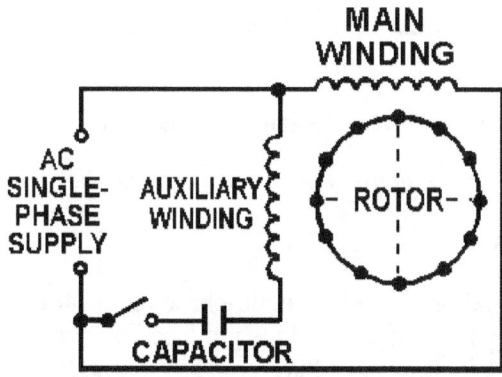

Figure 4-11.—Capacitor-start, ac induction motor.

When nearly full speed is obtained, a centrifugal device (the starting switch) cuts out the starting winding. The motor then runs as a plain single-phase induction motor. Since the auxiliary winding is only a light winding, the motor does not develop sufficient torque to start heavy loads. Split-phase motors, therefore, come only in small sizes.

RESISTANCE-START.—Another type of split-phase induction motor is the resistance-start motor. This motor also has a starting winding (shown in fig. 4-12) in addition to the main winding. It is switched in and out of the circuit just as it was in the capacitor-start motor. The starting winding is positioned at right angles to the main winding. The electrical phase shift between the currents in the two windings is obtained by making the impedance of the windings unequal. The main winding has a high inductance and a low resistance. The current, therefore, lags the voltage by a large angle. The starting winding is designed to have a fairly low inductance and a high resistance. Here the current lags the voltage by a smaller angle. For example, suppose the current in the main winding lags the voltage by 70°. The current in the auxiliary winding lags the voltage by 40°. The currents are, therefore, out of phase by 30°. The magnetic fields are out of phase by the same amount. Although the ideal angular phase difference is 90° for maximum starting torque, the 30-degree phase difference still generates a rotating field. This supplies enough torque to start the motor. When the motor comes up to speed, a speed-controlled switch disconnects the starting winding from the line, and the motor continues to run as an induction motor. The starting torque is not as great as it is in the capacitor-start.

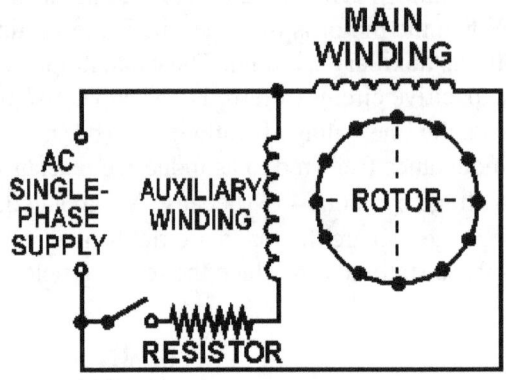

Figure 4-12.—Resistance-start ac induction motor.

Q12. How do split-phase induction motors become self-starting?

Shaded-Pole Induction Motors

The shaded-pole induction motor is another single-phase motor. It uses a unique method to start the rotor turning. The effect of a moving magnetic field is produced by constructing the stator in a special way. This motor has projecting pole pieces just like some dc motors. In addition, portions of the pole piece surfaces are surrounded by a copper strap called a shading coil. A pole piece with the strap in place is shown in figure 4-13. The strap causes the field to move back and forth across the face of the pole piece. Note the numbered sequence and points on the magnetization curve in the figure. As the alternating stator field starts increasing from zero (1), the lines of force expand across the face of the pole piece and cut through the strap. A voltage is induced in the strap. The current that results generates a field that opposes the cutting action (and decreases the strength) of the main field. This produces the following actions: As the field increases from zero to a maximum at 90° , a large portion of the magnetic lines of force are concentrated in the unshaded portion of the pole (1). At 90° the field reaches its maximum value. Since the lines of force have stopped expanding, no emf is induced in the strap, and no opposing magnetic field is generated. As a result, the main field is uniformly distributed across the pole (2). From 90° to 180° , the main field starts decreasing or collapsing inward. The field generated in the strap opposes the collapsing field. The effect is to concentrate the lines of force in the shaded portion of the pole face (3). You can see that from 0° to 180° , the main field has shifted across the pole face from the unshaded to the shaded portion. From 180° to 360° , the main field goes through the same change as it did from 0° to 180° ; however, it is now in the opposite direction (4). The direction of the field does not affect the way the shaded pole works. The motion of the field is the same during the second half-cycle as it was during the first half of the cycle.

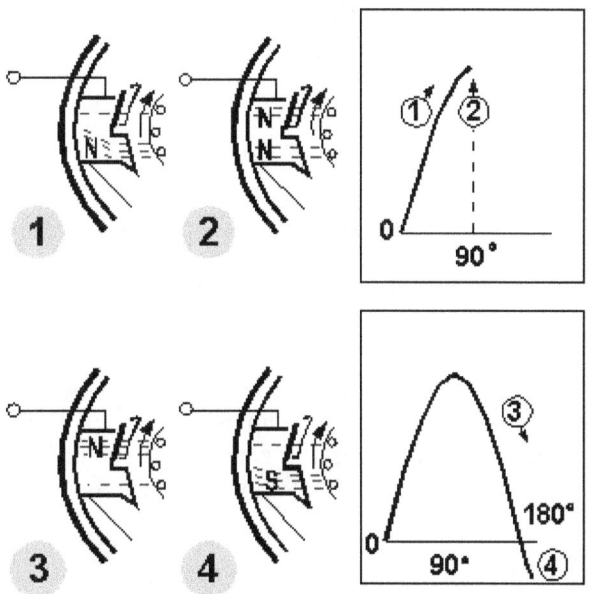

Figure 4-13.—Shaded poles as used in shaded-pole ac induction motors.

The motion of the field back and forth between shaded and unshaded portions produces a weak torque to start the motor. Because of the weak starting torque, shaded-pole motors are built only in small sizes. They drive such devices as fans, clocks, blowers, and electric razors.

Q13.   Why are shaded-pole motors used to drive only very small devices?

Speed of Single-Phase Induction Motors

The speed of induction motors is dependent on motor design. The synchronous speed (the speed at which the stator field rotates) is determined by the frequency of the input ac power and the number of poles in the stator. The greater the number of poles, the slower the synchronous speed. The higher the frequency of applied voltage, the higher the synchronous speed. Remember, however, that neither frequency nor number of poles are variables. They are both fixed by the manufacturer.

The relationship between poles, frequency, and synchronous speed is as follows:

$$n \text{ (rpm)} = \frac{120f}{p}$$

where n is the synchronous speed in rpm, f is the frequency of applied voltage in hertz, and p is the number of poles in the stator.

Let's use an example of a 4-pole motor, built to operate on 60 hertz. The synchronous speed is determined as follows:

$$n = \frac{120f}{p}$$

$$n = \frac{120 \times 60}{4}$$

$$n = 1800 \text{ rpm}$$

Common synchronous speeds for 60-hertz motors are 3600, 1800, 1200, and 900 rpm, depending on the number of poles in the original design.

As we have seen before, the rotor is never able to reach synchronous speed. If it did, there would be no voltage induced in the rotor. No torque would be developed. The motor would not operate. The difference between rotor speed and synchronous speed is called slip. The difference between these two speeds is not great. For example, a rotor speed of 3400 to 3500 rpm can be expected from a synchronous speed of 3600 rpm.

SUMMARY

This chapter introduced you to the basic principles concerning ac motors. While many variations of types exist, the three types presented provide you with background for further study if you require more extensive knowledge of the subject. The following information provides a summary of the major subjects of this chapter for your review.

The three **AC MOTOR TYPES** presented are the series, synchronous, and induction ac motors.

AC SERIES MOTORS are nearly identical to the dc series motors. Special construction techniques allow ac series motors to be used as **UNIVERSAL MOTORS**, operating on either ac or dc power.

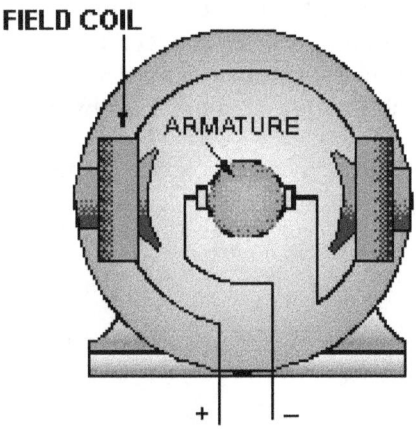

ROTATING FIELDS are developed by applying multiphase voltages to stator windings, which consist of multiple field coils. This rotating magnetic field causes the rotor to be pushed and pulled because of interaction between it and the rotor's own field.

TWO-PHASE ROTATING FIELDS require two pairs of field coils displaced by 90°. They must be energized by voltages that also have a phase displacement of 90°.

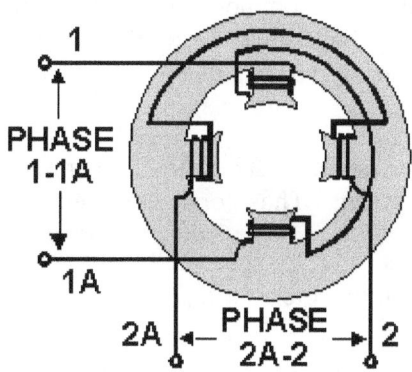

THREE-PHASE ROTATING FIELDS require three pairs of windings 120° apart, energized by voltages that also have a 120-degree phase displacement.

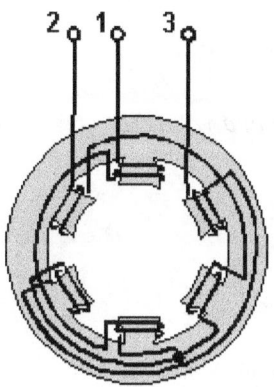

SYNCHRONOUS MOTORS are specifically designed to maintain constant speed, with the rotor synchronous to the rotating field. Synchronous motors require modification (such as squirrel-cage windings) to be self-starting.

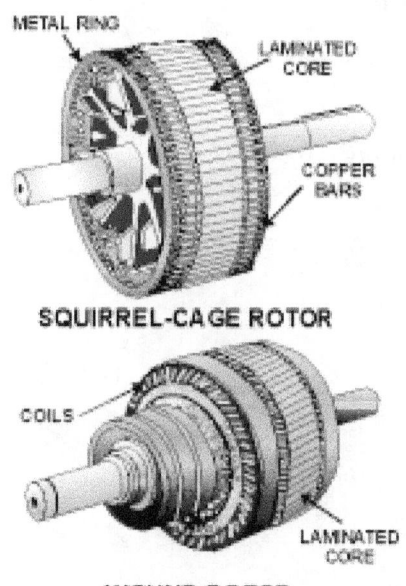

SQUIRREL-CAGE ROTOR

WOUND ROTOR

INDUCTION MOTORS are the most commonly used of all electric motors due to their simplicity and low cost. Induction motors may be single-phase or multiphase. They do not require electrical rotor connection. Split-phase motors with special starting windings, and shaded-pole motors, are types of single-phase induction motors.

SYNCHRONOUS SPEED is the speed of stator field rotation. It is determined by the number of poles and the frequency of the input voltage. Thus, for a given motor, synchronous speed is constant.

SLIP is the difference between actual rotor speed and the synchronous speed in induction motors. Slip must exist for there to be torque at the rotor shaft.

ANSWERS TO QUESTIONS Q1. THROUGH Q13.

A1.   *Series, synchronous, induction.*

A2.   *To power small appliances.*

A3.   *They operate on either ac or dc.*

A4.   *The number of phases in the applied voltage.*

A5.   *90°.*

A6.   *Number and location of field poles.*

A7.   *Constant speed required by some loads.*

A8.   *They are simple and inexpensive to make.*

A9.   *Slip.*

A10. *Load.*

A11. *Single-phase induction motor.*

A12. *By using combinations of inductance and capacitance to apply out-of phase currents in starting windings.*

A13. *They have very weak starting torques.*

www.ingramcontent.com/pod-product-compliance
Lightning Source LLC
Chambersburg PA
CBHW080619180526

45168CB00007B/2979